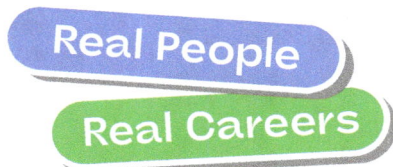 

ZOOKEEPER

© 2024 Julie Dascoli

All rights reserved. No part of this book may be reproduced or transmitted in any form or by any means, electronic or mechanical, including photocopying, recording or by any information storage and retrieval system, without prior permission in writing from the publisher.

Published in 2024 by Amba Press, Melbourne, Australia.
www.ambapress.com.au

Previously published in 2015 by Hawker Brownlow Education.
This edition replaces all previous editions.

ISBN: 9781923116849 (pbk)
ISBN: 9781923116856 (ebk)

A catalogue record for this book is available from the National Library of Australia.

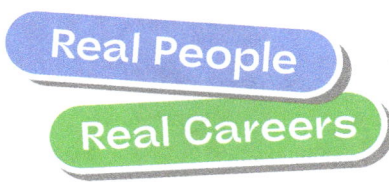

ZOOKEEPER

Written by Julie Dascoli

Photography by Laura Dascoli

Dear Reader,

Welcome to this volume of the *Real People Real Careers* series. I hope you'll enjoy learning about another exciting job people can do.

Before you read on, I'd like to say a few thank-yous to the people who helped to make this book possible.

Firstly, thank you to Laura Dascoli, who took the photographs you see in the book, and to Donna Dascoli, who provided initial editing and computer support services.

Secondly, my thanks to the staff and students in Years 4, 5 and 6 of the Mossgiel Park Primary School class of 2016 for their unwavering help and support.

And finally, I'm doubly grateful to Mike, who generously gave up his time to help others learn about his profession – and to show them all the ways in which his job rules!

Happy reading!

Julie Dascoli

ZOOKEEPER

My name is Michael, but everyone calls me Mike. I am a **Zookeeper**.

I have been an animal lover for as long as I can remember. As a child I was always collecting and playing with different kinds of insects, and bringing them home with me.

Reptiles have always been my favourite. I used to read about them, watch **documentaries** about them, and have always had an enormous curiosity about reptiles. I still have eight lizards as pets at home. I have a pair of Angle-Headed Rainforest Dragons, three Pink-Tongue Skinks, a pair of Knob-Tailed Geckos and a Bearded Dragon.

I attended local state primary and secondary schools. At high school I took subjects such as English, Science and Maths, but **Biology** was always my favourite.

On completing Secondary School, I got a job as a **Laboratory Technician**. While I was there, I was working with a lady who was **volunteering** at a huge, well-known **sanctuary** located just out of Melbourne. This lady encouraged me to do the same.

Straight away I approached the sanctuary regarding becoming a volunteer, and I started immediately.

This was amazing. I was so happy to be surrounded by animals and, after three months, as I had some previous experience in handling reptiles I was offered a job with the sanctuary.

Once I was settled in my new job, I embarked upon some more specific study to further my skills and knowledge.

I obtained a Certificate of Applied Science and a Certificate of Zookeeping. Zookeepers look after all different kinds of animals at zoos and sanctuaries. I have chosen reptiles as my speciality.

> Reptiles have always been my favourite. I used to read about them, watch documentaries about them, and have always had an enormous curiosity about reptiles.

Tasks I perform every day

- → cleaning the surrounds of the **Reptile House**
- → clean the windows of the **enclosures**
- → while doing the above tasks, I check on the animals in exhibits
- → observing the animals to ensure that they are healthy
- → cleaning the exhibits
- → feeding the animals. The animals are all on a personal feeding programs. For example, the larger **Pythons** may be fed two large rats and this will be all they need for a month. The smaller snakes and lizards may get a little bit every day. It is my job to follow the individual programs to ensure **optimal health** for the reptiles I care for.

- engaging in visitor experiences. Most days I bring our **Lace Monitor,** called Chantilly, out the front of the reptile centre and allow visitors to pat him and ask questions at a certain time. I also supervise groups to tour the area where we keep various examples of snakes and lizards.

- do paperwork. Everything is recorded in the reptile centre. Births, deaths and illnesses, eggs in the incubator, everything must be recorded. This is how we learn more and more.

- planning ahead for school visits and tours.

Protective sleeve

'Man Down' radio

Water for the reptiles

Interesting facts about my job

- → My job is both my hobby and my employment.
- → I get to bring the reptiles out of the enclosures to show our visitors and talk to them about the animal.
- → I work 8 hours per day.
- → I have 1 hour for lunch.
- → I have two 15-minute tea breaks per day.
- → I have a communication radio with me at all times. This radio is programmed so that if I am lying down due to injury it will automatically alert someone to come to my aid.

- My favourite task is to get 'Chantilly', our resident Lace Monitor, and show him to the visitors.
- My least favourite task is going to meetings.

Snakebite kit

Tool for hooking snakes

'Chantilly' the Lace Monitor

An incubator

I wear a complete uniform to work every day with the logo for the sanctuary on it. This includes steel-capped safety boots, long pants in winter and shorts in summer. I wear a shirt with a logo all year round, and a jacket in the winter. While I am in the reptile enclosure I take my jacket off, as it is temperature-controlled for the welfare of the cold-blooded animals but quite hot for humans.

As a zookeeper, wearing a uniform each day is incredibly practical and unifying. Our uniforms are designed for durability and comfort, essential for the physical nature of our work, from cleaning enclosures to feeding animals. They also create a sense of team spirit and professionalism, ensuring visitors can easily identify us if they have questions or need assistance.

Utensils for preparing food

You can do my job if you:

- → love animals and care for their **preservation** and **conservation**
- → have a passion to educate people about animals
- → have good communication skills
- → don't mind if you get dirty
- → enjoy public speaking
- → enjoy working with young people
- → enjoy being outdoors

Reptile food

Associated occupations

- → Research Scientists
- → Zookeepers specialising in all other animals
- → **Park Rangers**
- → **Marine Biologists**
- → **Zoologist**
- → Wildlife Biologist
- → Ecologist
- → Animal Trainer or Handler
- → Aquarist

> As a zookeeper, I can't imagine doing any other job; working with reptiles has always been my dream and my passion.

Postscript

Mike continues to work at this famous and very popular sanctuary. He loves caring for the reptiles and teaching the public about them. He especially enjoys helping those who are scared of reptiles to understand them better and show them that they are not as scary as they thought.

This is Mike's dream job!

Glossary

Biology	The science of life and living organisms. *Biology was Mike's favourite subject at school.*
Conservation	In the sanctuary this is the protection of plants, animals and environment. *Mike is committed to conservation and preservation within his job.*
Documentaries	A film, television show or radio show that provides a factual report on a particular subject. *Mike has enjoyed watching documentaries about animals all his life.*
Enclosure	An area that is surrounded by walls or fences and, in the case of the reptile house, wood and glass. *A part of Mike's job is to keep the inside and outside of the reptile enclosure clean.*
Incubator	A device that is used to keep eggs warm before they hatch. *Mike looks after the incubators when they have reptile eggs in them.*
Laboratory technician	The person who does the hands-on work in the laboratory under the guidance of a scientist. *Mike's first job after high school was as a laboratory technician.*

Lace Monitor	A very large lizard, also called a Goanna or Lace Goanna. They can grow from 1.5 to 2 metres long. *The sanctuary has a Lace Monitor called Chantilly. Mike brings her out to show the visitors and answer questions about her.*
Marine biologist	A scientist committed to studying all things under the sea and other waterways.
Optimal health	To be in the best health that is possible. *Mike keeps a close watch over the reptiles to ensure they stay in optimum health at all times.*
Park ranger	A person entrusted with protecting and preserving park-lands, national parks and everything in them.
Preservation	To keep something safe or in its original state. *A part of Mike's job is to help maintain the preservation of the animals and their living environment.*
Python	A large, heavy-bodied, non-venomous snake that is found in tropical climates. *Mike looks after the pythons at the sanctuary.*
Reptile house	An enclosure where reptiles are kept. *Most of Mike's work is in the Reptile house.*

Reptiles	Cold-blooded animals including snakes, lizards, turtles and tortoises. They typically have dry scaly skin and lay soft-shelled eggs on land. *Mike is qualified to work with all animals, however he specialises in reptiles.*
Sanctuary	A nature reserve where flora and fauna are conserved and studied. *Mike volunteered and works at a very famous sanctuary out of Melbourne.*
Speciality	An area of study to which one has devoted their time and skill to become an expert on. *Mike chose reptiles as his speciality.*
Volunteering	To volunteer is to do work for an organisation without getting paid. *Mike did some volunteering at the sanctuary so that he could work with the animals even though he was not getting paid. This gave him a chance to display his knowledge and work ethic, and ultimately got him a job.*
Zookeeper	An animal attendant employed at a zoo or sanctuary. *Mike's job title is Zookeeper.*
Zoologist	A scientist that studies animals. Zoologists are experts on everything about animals — from their cells to their evolution.

Other titles in this series

www.ingramcontent.com/pod-product-compliance
Lightning Source LLC
Chambersburg PA
CBHW070343120526
44590CB00017B/3002